My Hands
Loretta Lang

Whenever I grow weary in teaching

Listening to "I can't do it" on point

Followed by the notion "I don't get it"

I know that like a wad of clay

Students' thinking can be molded in my hands

Being better prepared is my duty

Answering questions I think they might ask

Accepting right and wrong answers

I see the motivation lit

Students' thinking can be molded in my hands

Dedicated to Superintendent Wanda Bamberg and the Aldine Independent School District

ABSTRACT

The idea to incorporate higher-order thinking (questioning, reasoning, strategies, activities, i.e.) into lessons, sometimes falls short of its delivery. Teachers struggle to consistently add these elements in their instructional practices.   In light of the new state assessment, considerations for teaching strategies, key words, and the like are far from the expected rigor built into the test. When approaching math from a science perspective (math is a science) students will use tools to explore, discover, and analyze mathematical outcomes.

Contents

Introduction

In math classes across America, students struggle to understand basic mathematical concepts. Unfortunately, if students do not understand these concepts in the lower grades, they incur shortfalls in upper-grade level skills. Moreover, due to the mandates of "No Child Left Behind" (NCLB), teachers negotiate between teaching to the test and structuring lessons that require more hands-on, investigative, and project-oriented tasks. Online NewsHour, a subsidiary news group of Public Broadcasting Service (PBS) investigated the effects of NCLB by interviewing teachers and administrators across America. Kathleen Smith, a veteran teacher from Illinois says, "Out went discovery learning, out went project based learning, out went open-ended problem solving. In came multiple choice warm-ups, quarterly 'chunking' tests, and 15 days of test practice. I was no longer a teacher; I was a test-prep coach"(2005).

Unlike Kathleen, don't park your instructional know-how on "Frustration Boulevard". Teaching and learning can be meaningful to encourage cognition (thinking to know) to metacognition (knowing to know).

Concepts vs. Strategies

Many students have mastered the idea of solving word problems by using strategies. Needless to say, strategies (key words and phrases) work during testing time; however, in real life can students problem solve without these clues? For example:

 Question #1

Mark had shirts in his closet and none of them fit. If he wanted to compare the 8 shirts to 12 new T-shirts, how many T-shirts to shirts could he not compare?

Note: If students have not learned the comparison concept, then this word problem may present a challenge to them. In most cases, this question would have probably been written in this manner (see question #1a).

Mark had 8 shirts in his closet and none of them fit. If he wanted to replace the 8 shirts with 12 T-shirts, how many more T-shirts than shirts would Mark have?

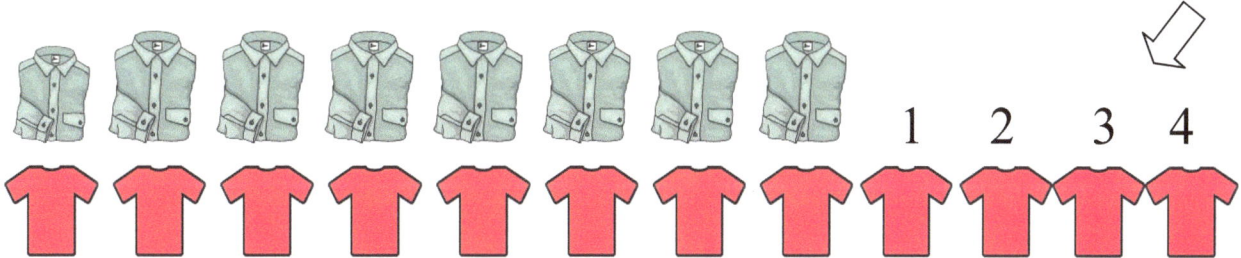

When comparing objects, numbers, i.e., the student should look for the difference.

Findings

The National Mathematics Advisory Panel/Final Report suggests:

> "Explicit instruction with students who have mathematical difficulties has shown consistently positive effects on performance with word problems and computation. Results are consistent for students with learning disabilities, as well as other students who perform in the lowest third of a typical class. By the term *explicit instruction*, the Panel means that teachers provide clear models for solving a problem type using an array of examples, that students receive extensive practice in use of newly learned strategies and skills, that students are provided with opportunities to think aloud (i.e., talk through the decisions they make and the steps they take), and that students are provided with extensive feedback (2008)."

Mathematics as a Science

Without a doubt, science experiments require the use of tools. If math is under the auspices of science, shouldn't students use tools when exploring, discovering or analyzing mathematical processes? The question is, are students familiar with the **underlying processes and mathematical tools** to assist them? Sound familiar? The Texas Education Agency (TEA) emphasizes the use of mathematical "processes and tools" in Chapter 111. Texas Essential Knowledge and Skills for Mathematics. What does this mean? In essence, students must know and apply the steps (plan) to solve mathematical problems/situations/experiences using tools (charts, tables, graphs, number lines, objects, manipulatives, technologies, i.e.). According to TEA, **"underlying processes and mathematical tools"** appears in connection with other skills. "These skills will not be listed under a separate reporting category. Instead, they will be incorporated into at least 75% of the test questions in reporting categories 1–5 and will be identified along with content standards (TEA, 2012)".

I like the idea that students can create their own tools to assist them mathematically. Unfortunately, time and time again, I see more strategies (underlining the question, circling the important numbers, identifying key words, i.e.) than the use of mathematical "processes and tools". Since underlying processes and mathematical tools are connected and identified with other skills, shouldn't it be connected to most of the instruction and learning experiences? Chapter 111. Texas Essential Knowledge and Skills for Mathematics states:

> "The process standards describe ways in which students are expected to engage in the content. The placement of the process standards at the beginning of the knowledge and skills listed for each grade and course is ***intentional***. The process standards weave the other knowledge and skills together so that students may be successful problem solvers and use mathematics efficiently and effectively in daily life. The process standards are integrated at every grade level and course (TEA, 2012)."

Exploration and Discovery!

I believe most math teachers in America would agree that students need to learn their mathematical facts. However, in considering mathematics as a science, asking students to "recall 7X9" does not allow them to discover a pattern, order, or predict an answer! When math is approached as a science, a better way to ask this question may be, "How is 3x3, 5x9, and 7x3 related"? This type of questioning will cause students to "think deeper".

Below are some examples of how students might respond to the teacher's question, "How is 3x3, 5x9, and 7x3 related":

Response 1: **All are times tables**
Response 2: **Some factors are greater than the other**
Response 3: **Each will have a product**

In observing these responses, one would say that the question was not fully answered. Well, always remember that there are no wrong answers-only responses by individuals who respond according to where they are in their thinking. Therefore, it is the teacher's responsibility to prepare lessons that connect to students' prior knowledge with new concepts.

Science Related Questioning

What if the teacher trained the students to look for a pattern and/or predictability concerning the aforementioned question? Then, the students' responses probably would have been different.

Factor		Factor	Product
3		3	9
5		9	45
7		3	21

Dividend		Divisor	Quotient
9		3	3
45		9	5
21		3	7

When conducting division with the same numbers the quotients will be odd

Have the students work with a partner to complete the table below.

Factors	Array	Even Product	Odd Product	Findings
3X3	@ @ @ @ @ @ @ @ @		X	**What did you discover about the products for the first set and second set of examples?** When two factors are odd and multiplied the product is odd. On the other hand, when two factors are even and multiplied the product is even.
5X9	@@@@@@@@@ @@@@@@@@@ @@@@@@@@@ @@@@@@@@@ @@@@@@@@@		X	
7X3	@@@@@@@ @@@@@@@ @@@@@@@		X	
2X2	@ @ @ @	X		**What will happen if 24 was divided by an even number?** _____ _____
2X8	@@@@@@@@ @@@@@@@@	X		
4X4	@@@@ @@@@ @@@@ @@@@	X		

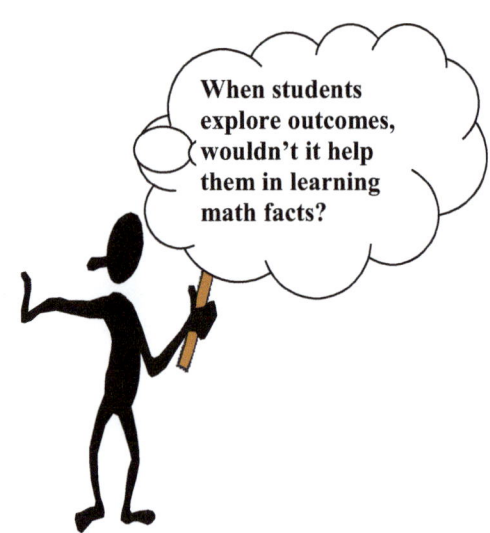

When students explore outcomes, wouldn't it help them in learning math facts?

5

If students could learn **two** major concepts to analyze and explore mathematical outcomes, wouldn't math seem simpler?

There are representations in everyday life that reflect an increase, decrease or equal outcome. Even though there are three concepts to this matrix, students who struggle with mathematical concepts usually do so because they do not recognize an outcome of an increase or decrease.

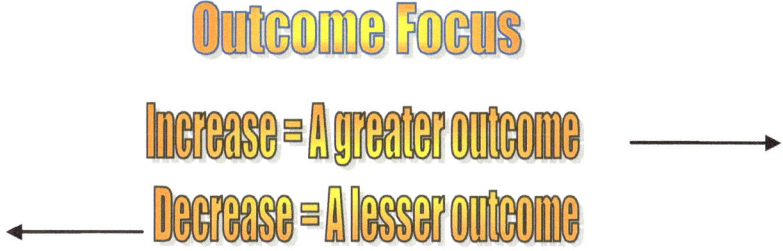

Outcome Focus

Increase = A greater outcome →

← **Decrease = A lesser outcome**

(Examples Below)

A mountain from its base to summit reflects a decrease to an increase in altitude. However, from its summit to the base would reflect the opposite-increase to decrease in altitude.

Mountain A

Summit/Top (increase) 20,000 feet	**Summit/Top (increase) 20,000 feet**

Base/bottom (decrease) 3,000 feet	**Base/bottom (decrease) 3,000 feet**

B Circumference

The Circumference of a pear can reflect an increase to decrease or decrease to increase-depending on what part is considered initially.

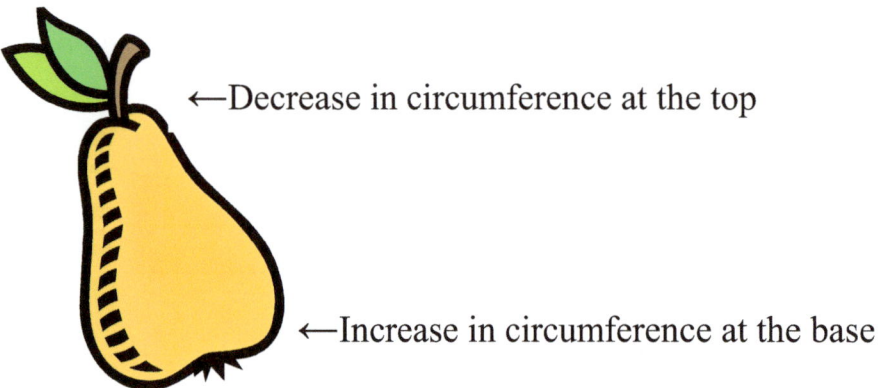

←Decrease in circumference at the top

←Increase in circumference at the base

C Hopscotch Game

Increase to Decrease   **Decrease to Increase**

D Trail

(From Point A to Point B there is a decrease to increase in width)

(From Point B to Point A there is an increase to a decrease in width)

E Number Representation

Decrease   Increase

Understanding Increase and Decrease Concepts

Summit/Top (increase) 20,000 feet

Summit/Top (increase) 20,000 feet

Base/bottom (decrease) 3,000 feet

Base/bottom (decrease) 3,000 feet

Note: Students tend to add the numbers before considering the increments by which the numbers increase or decrease (from one number to the other). To add the numbers all together would mean that after 20,000 feet there is an additional consideration of 3,000 feet.

9

Back to the Basics

If you ask a student what is the relationship between 2 and 6, they would have to think about it. Knowing the distance between numbers, and the increase and decrease concept can help students develop a keen sense of mathematical direction (increase + or X) and (decrease – or /) and rule out any misconceptions.

The Number Line

As a special education inclusion teacher, I am in and out of different classes. Often I see students struggling with mathematical outcomes because they do not utilize visual tools that can assist them in analyzing a mathematical problem properly. I believe that one of the greatest tools that students can utilize that would offer them analytical clarity is the number line. With the number line, students can order numbers, analyze number patterns, create fractions, determine number proximity, factor numbers, i.e.

Outcome Model Tool

All of the students I serve have accommodations. These students are afforded these tools as a way to increase their academic success. I assist regular and special education students and often the regular education students ask me why they can't use some of the tools when taking their test. Usually my answer is never sufficient especially when I know about their academic struggles. With this in mind, I believe that God gave me a tool that any student could use/create when completing assignments, projects, homework, teacher-made, district and/ or state tests while still being in compliance with state mandates. It is called the *Outcome Model Tool*. This tool's major feature is a number line focusing on increase and decrease possibilities with built-in aids. With the facilitation of this tool, students can plan, analyze, investigate, discover, confirm, construct, and process mathematical outcomes.

Outcome Model Tool
(Focusing on Increase-Decrease Concepts)

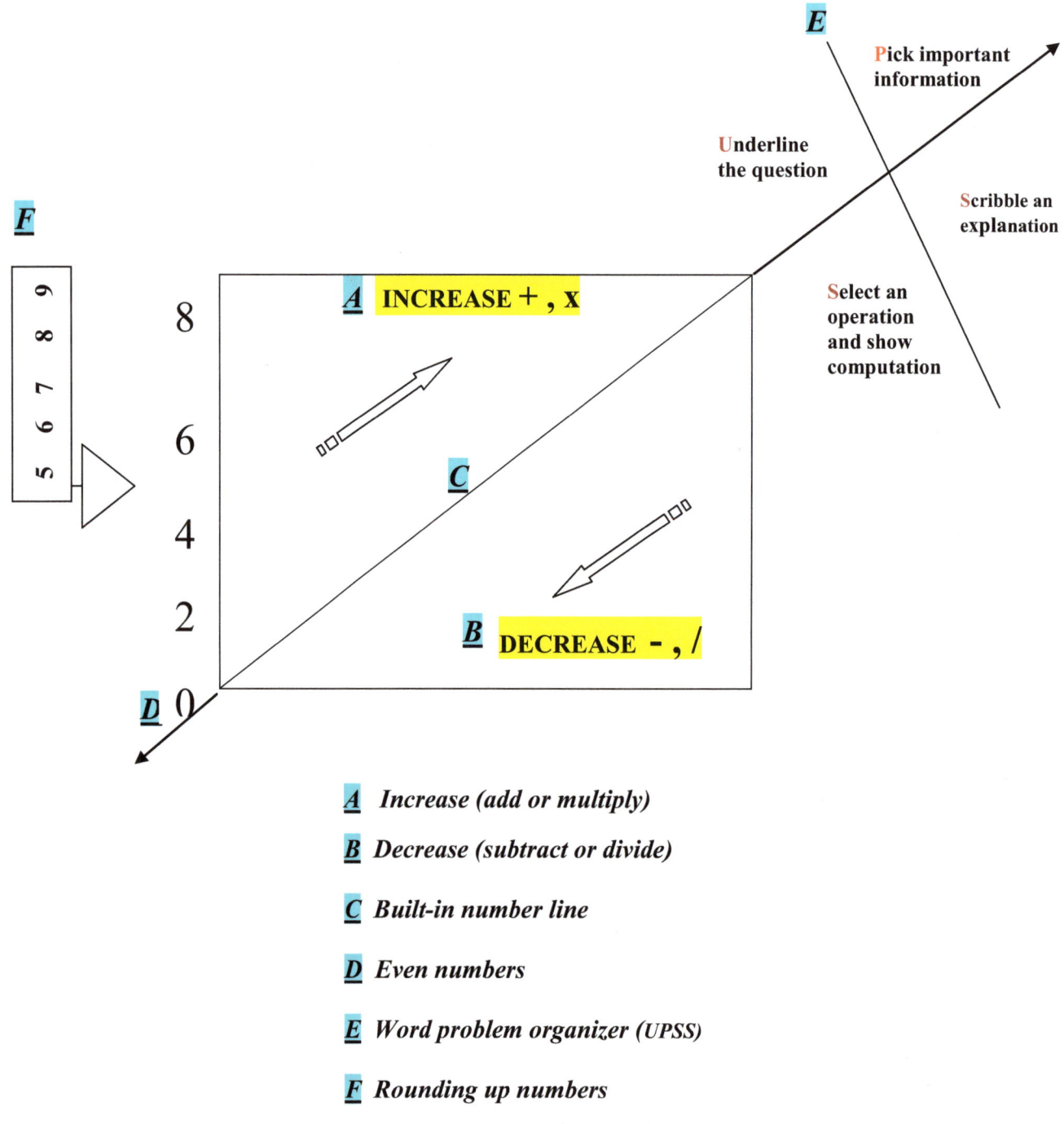

A *Increase (add or multiply)*

B *Decrease (subtract or divide)*

C *Built-in number line*

D *Even numbers*

E *Word problem organizer (UPSS)*

F *Rounding up numbers*

Using the tool

Students do not have to think *long and hard* before choosing an operation. In fact, the options are clear. When expecting an outcome of more, the increase operations (+, x) should be utilized. However, when expecting an outcome of less, students should use the decrease operation (- , /) options.

 **The Big Question**

Do you have to use all of the model's components at one time? No. Students will add the different components as they become more familiar with the tool. For example, students in kindergarten may only use components A & B.
(See below)

⭐**Question #1**

Jose had 1 ball and his mother gave him 3 balls. How many balls does Jose have now?

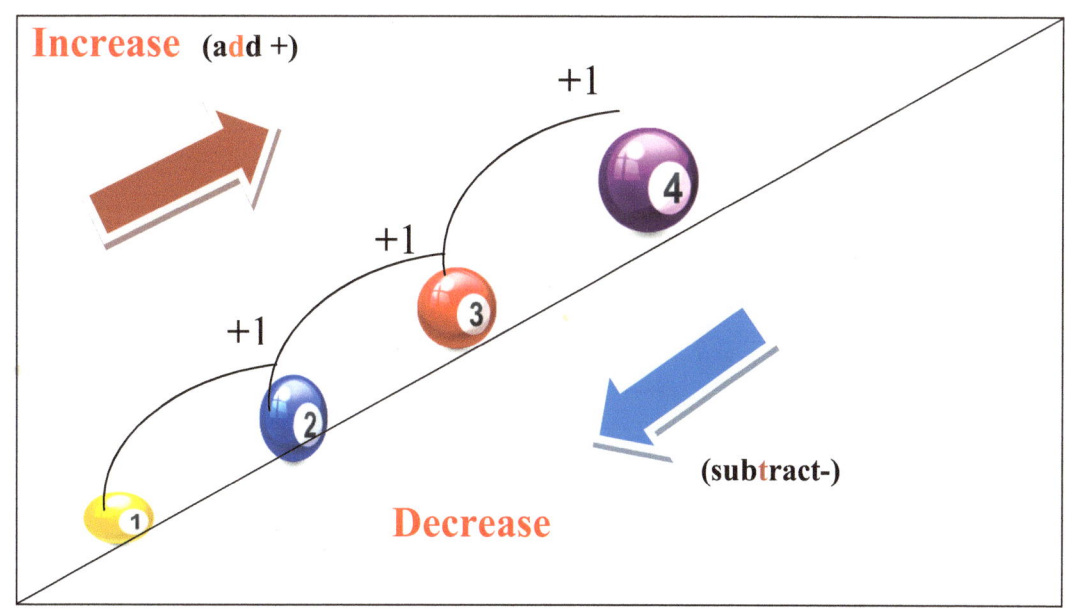

Answer: Now, Jose has four balls ⟶ $1 + 3 = 4$

 **Question #2**

On Monday Linda noticed that 154 people visited her mom's coffee shop before 6:00 a.m. each day. By evening the number seemed to triple. **About how many people visited the coffee shop by the evening?**

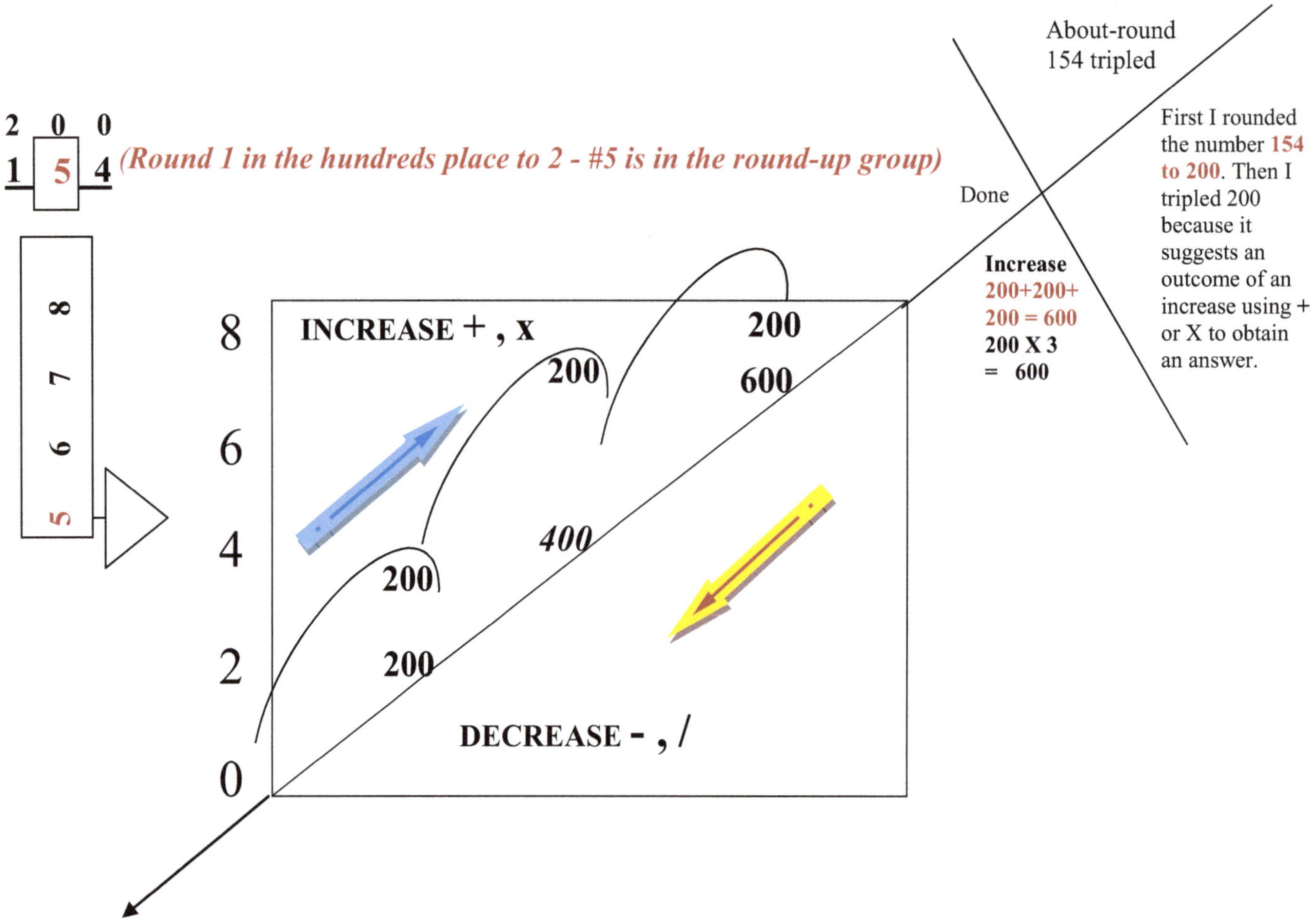

2 0 0

1 5 4 *(Round 1 in the hundreds place to 2 - #5 is in the round-up group)*

About-round
154 tripled

First I rounded
the number **154**
to **200**. Then I
tripled 200
because it
suggests an
outcome of an
increase using +
or X to obtain
an answer.

Done

Increase
200+200+
200 = 600
200 X 3
= 600

8 7 6 5

8 **INCREASE + , X**   **200**

200   **600**

6

4   *400*

200

2   **200**

DECREASE - , /

0

Manipulating the tool

The more students use this tool the more comfortable they will be in displaying and performing different operations to solve mathematical problems. For example, instead of using the x axis for even numbers, this axis can be used as a number line.

 **Question #3**

There are 10 houses on Orville Street. Rachel lives in the third house on the block and Lisa's house is two doors before Michael's house. Michael lives in the ninth house on the block. Where would Lisa's house be located?

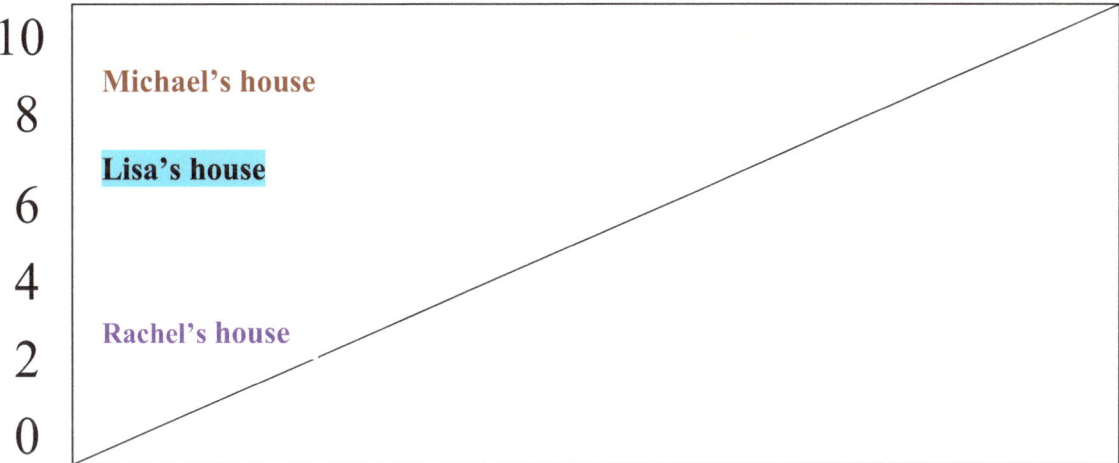

Bar Graphs

 **Question #4**

Smith Academy had a race for class president. Debbie had more votes than Lisa and Lisa had fewer votes than Mike. Rachel won as class president and Cindy had the least amount of votes than everyone. Graph the results below using your own numbers on the x and y axis.

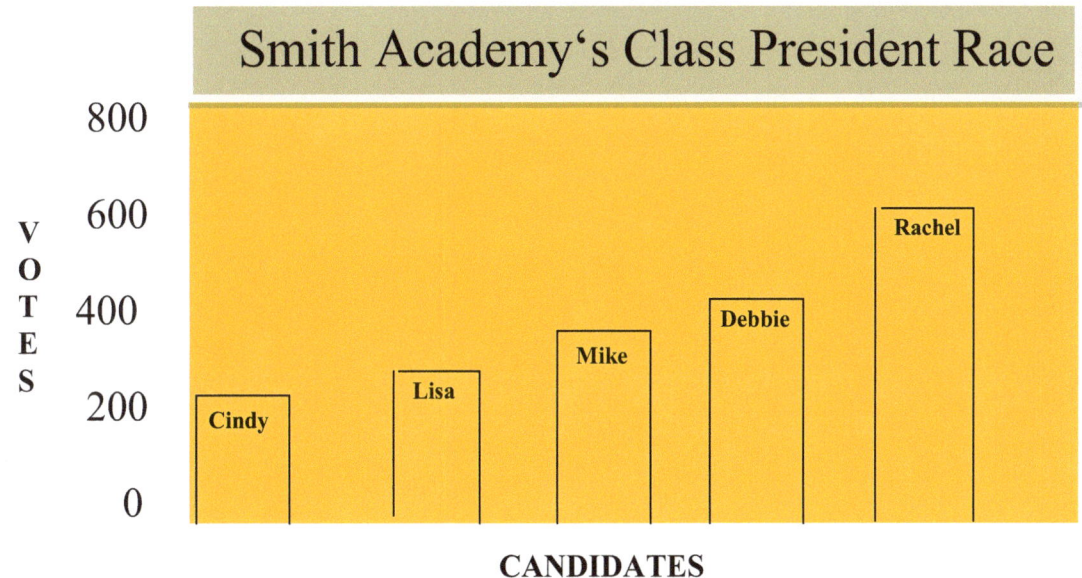

Question #5

Mary had a party. She wanted to buy 7 party favors for each of her friends. If she invited four people, how many party favors would she have to buy for 4 people?

Input	Output
People	Party Favors
1	7
2	14
3	21
4	?

Have the students create the Outcome Model Tool to show how the numbers have increased. (Refer to the example below)

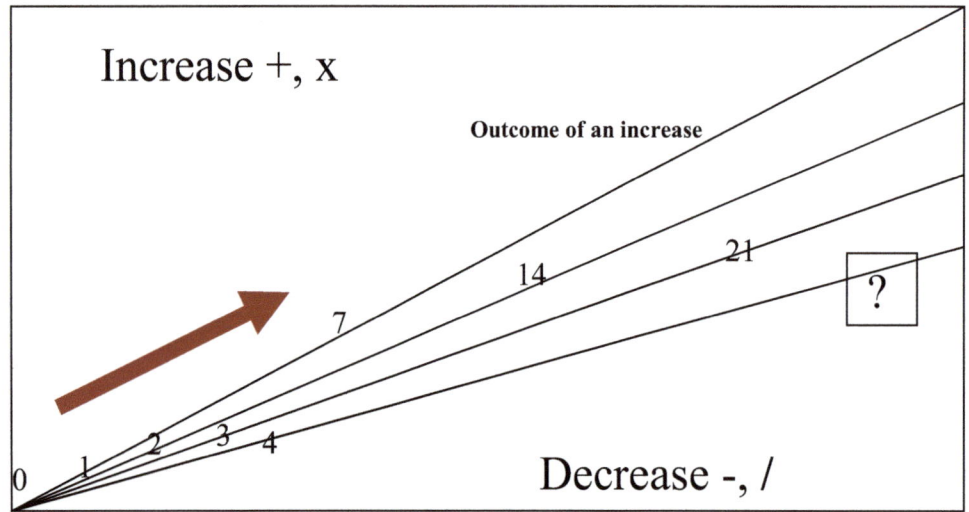

Creative Usage

As students use the tool, they will start to see that the process starts with a simple rectangle or square. They can draw, calculate, determine possibilities or do whatever is necessary to devise a plan!

**Question #6**

There is a party with 24 people. If ¾ of the people rely on public transportation, how many people will leave the party ___not___ using public transportation?

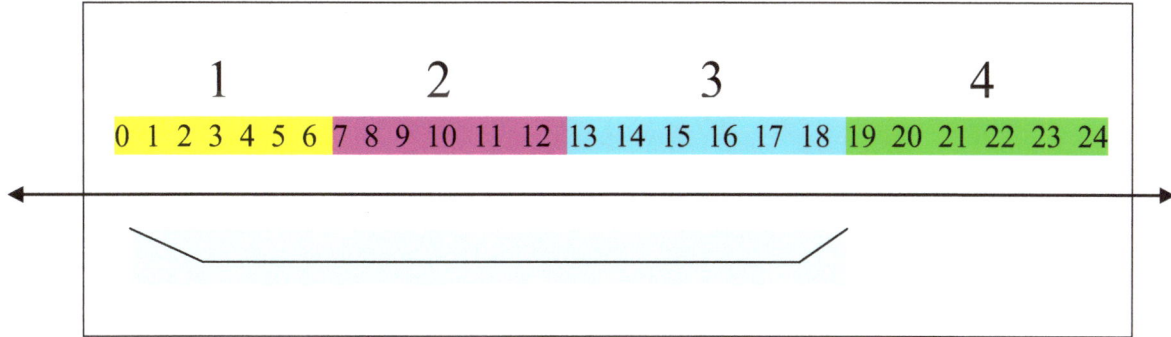

Students must understand that the denominator acts as the divisor. When 24 is divided by 4 it equals 6. So, every sixth person would represent a fourth.

Question #7

Jaime wants to decide how to cut his rectangular cake. He asks his friends if they want 1/3, 1/5, or 1/7 of his cake. Rasheed said that he wanted a large piece so he chose 1/7 over the other fractional choices. Is Rasheed correct in assuming that he would get a larger piece when the cake is cut into 7ths? *(Provide a drawing to show reasonableness)*

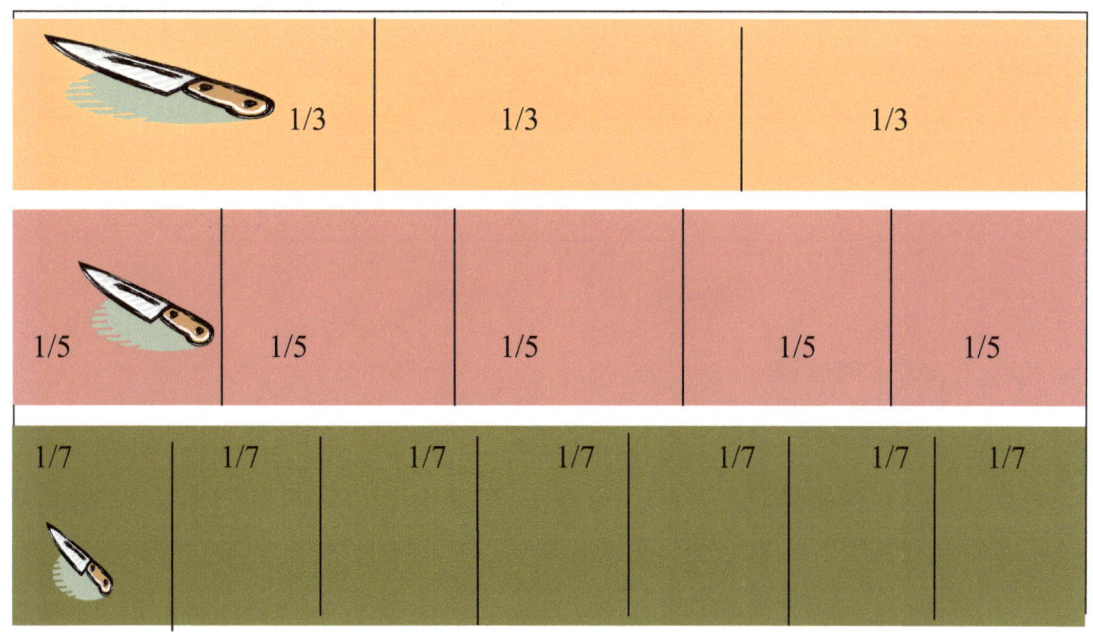

Geometry

Students can use the matrix below to develop roads, communities, parks, i.e.
(Notice the angles and lines within the matrix - use, duplicate, and/or erase lines to draw)

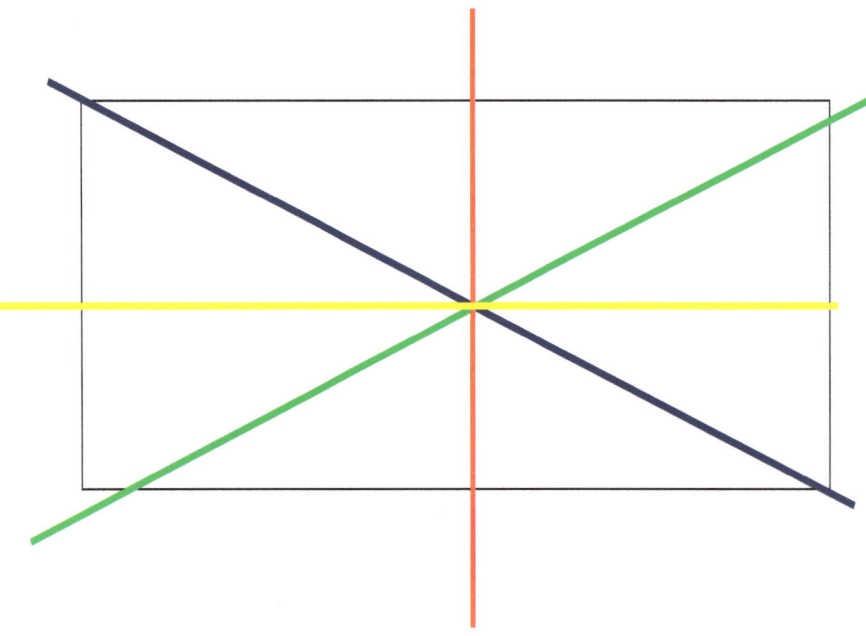

Assignment:

You are developing roads to access a new neighborhood park. Create two roads that will run parallel to access the park. Include and label a petting zoo, golf course, parking lot, and fountain. Make sure that there are three acute angles that share the same vertex.

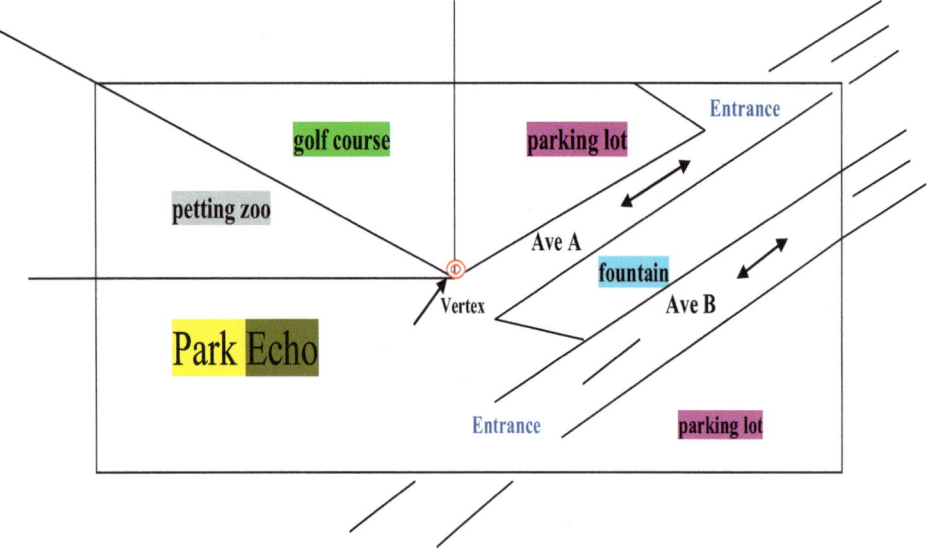

Endless Possibilities

There are endless possibilities in using the Outcome Model Tool! By introducing this tool, I believe that students will understand that mathematics requires the use of some type of tool to satisfy mathematical quarries. Even if students do not use this particular tool, my hope is that they create their own mechanism to integrate the skill, ***underlying processes and mathematical tools*** into their mathematical experiences.

Summary

As an educator, I have seen students struggle to understand basic mathematical concepts. Thankfully, there is hope for every student to become great mathematicians! Understanding concepts and when and how to use the proper tool are key essentials. Even as I pen these words, I realize that some students have not "mastered" the thought that there are predictable outcomes in math-math is a science. The operations begin with addition, subtraction, multiplication, and division. All mathematical answers/outcomes will show an increase, decrease or equality (ratios, estimations, fractions, integers, whole numbers, percentages, i.e.).   Just knowing this will put the students on the right track and remove some of the "fogginess" that has plagued their mathematical thinking.

On the other hand, teachers no longer have to struggle to incorporate higher-order thinking into their instructional practices. It will come naturally-when looking at math from a science perspective. Like science, math is predictable, shows patterns, and causes one to look a little deeper to find the right answer. Maybe anyone could look at 7x7 and say "49" but discovering that odd factors always produce an odd product fosters a better understanding of how math is predictable-a science!

References

National Mathematics Advisory Panel. *Foundations for Success: The Final Report of the National Mathematics Advisory Panel*, U.S. Department of Education: Washington, DC, 2008.

Thompson, C. *No Child Left Behind: Tales from the Frontlines.* Public Broadcasting Service Online News Hour Archives. August 21, 2005. MacNeil/Lehrer Productions. .http://www.pbs.org/newshour/indepth_coverage/education/no_child/frontlines.html

Texas Education Agency. *Texas Essential Knowledge and Skills for Mathematics.* Texas Administrative Code, Title 19, Part II, Chapter 111. 2012. http://ritter.tea.state.tx.us/rules/tac/chapter111/ch111a.html

About The Author

Loretta Jefferson-Lang is an educator in the Aldine Independent School District and lives in Houston, Texas with her husband, mother, son, and beloved pooch, Sophia. She currently holds a Bachelor of Art's Degree in Music from Xavier University, an MBA from Devry University, and a Master's Degree from the University of Phoenix in Education Administration and Supervision.

Loretta has served in a teaching capacity for 30 years. Currently, she assumes the responsibility as a Smith Academy special education teacher and 3rd and 4th grade math facilitator.

Through the years, she continually noticed common deficits in students' basic understanding of mathematics and began free tutorial services for struggling students at school and in her neighborhood. The students were introduced to the increase and decrease concepts while utilizing the "Outcome Model Tool". Soon, the students began to make remarkable progress and surpass all expectations. With such success, the desire to reach more students, teachers, and parents became a major focal point.

Notes

www.ingramcontent.com/pod-product-compliance
Lightning Source LLC
Chambersburg PA
CBHW041307180526
45172CB00003B/1012